Know about Nanotechnology
By
Chakrapani Srinivasa

Know about Nanotechnology
By
Chakrapani Srinivasa
Copyright 2020 Chakrapani Srinivasa

Dedicated to my dear parents

About the Author

Chakrapani Srinivasa (Padmaja), Freelance journalist from India possesses Bachelor degree in Engineering (B.E) and Post graduate in Business Management (MBA) with Distinction. He has worked as Associate Editor of 'Naradar' fortnightly journal in Chennai, India. He is the Senior Editor of the journal "The Divineness".

Contributed articles, short stories and travelogues in leading journals like Ananda Vikatan, Kumudam, Savi, Kalki, Dinamani Kadhir, Dinamani daily, Idhayam Pesukirathu, Naradar etc

He has written articles and e books through Smashwords Inc, Kindle Direct Publishing, Atlanta publications, Cooperjal publications (UK), lulu.com, ezinearticles.com, shvoong.com, iproclaim.com (USA) and TCC news (Germany).

He is the Consulting Editor: Contemporary Who's Who-Research Board of Advisers of ABI, USA.

Preface

Nano technology is nothing but knowing and control of matter at dimensions of 1 to 100nm, which enable it to use for sophisticated applications. A nanometer is 10^9 of a meter. Nano technology comprises imaging, measuring, modeling and manipulating the matter, with this length of scale. In this level, the physical, chemical and biological behavior of material will vary in fundamental and valuable ways from both the properties of individual atoms and molecules or bulk matter. Hence, Nano technology experts are engaged in creating better materials and devices with unique systems that exploit these new properties.

Role of Nano Technology in India

Regional Centre for Education and Training in Bio Technology is an ideal contribution for promoting Nano-biotechnology. Many applications for Defense needs have been brought to lime light by them.

Vehicles and Nano

Nano technology plays a vital role in defense in various aspects.

Engine condition sensors to ensure safety for defense vehicles is an important contribution from Nano technology.

Long range Sensors, Comfort Sensors, Smart Tyre with Nano Composites, Self adaptive of Skin and Structures, High Temperature Plastics, Oil and Fuel Sensors, Nano Tube Shock Absorbers, Reactive Nano Armor, Fuel Cells, Bio Fuels, Hybrid Power, Droplet size Injections, Oil and Fuel Quality Sensors, Nano Catalysts and Nano Membranes, Bio-Robotics, Gyros Navigation, 1D tags, Long Range Guidance Sensor, Surveillance Sensors. Position and Motion Sensors are valuable contributions from Nano technology to serve the warriors.

Soldiers and Nano

Brain machine interface, Exo Skeletons, 360° Adaptive Vision Systems, Ambient Intelligence, Bio-metric identification, Event driven info, Digital Identification, Position and Motion Sensors, Smart Textiles, Lightweight protective Clothes, Wearable rf/ Electric/ Opto/ Accoustible Health Sensors, Bio Fluidic Sensors, Tissues Engineering, Active Biochemical protections, Active Bio Chemical protection, Artificial Organs and Blood, Wireless Sensor Implants, Nerve and Muscle Stimulation, Energy Scavenging, Solar Cell in Foil, Wearable Power, Membranes food / Water and Air, PDA Tags, Flexible Displays, Bio Fluidic Sensors, Digital Identification and Nuclear Battery are great contributions of nano for defense.

Institutions and Defense

14 great institutions function under Centre of Excellence (COE) in 19 centers in India. Along with Department of Science and Technology we have DBT Department of Bio-technology to execute Nano technology and Life Sciences.

Council of Scientific and Industrial Research has a big team of Scientist trained in Russia, U.K. U.S.A and Israel to focus attention on needs of Naval, Air Force and Army men. It has 38 Laboratories to coordinate with DRDO and defense organizations.

Science and Engineering Research Council do a yeomen service to Nano technological developments in India. They have links and assistance from Taiwan, Japan, USA, Russia, S.Korea, South Africa and Brazil.

A MOU has also been signed between UNESCO and India to conduct R & D works in Nano Technology and training to our budding scientists in various institutes like IIT in Mumbai, Delhi and Kharagpur.

Regional Centre for Education and Training in Bio Technology is an ideal contribution for promoting Nano-biotechnology. Many applications for Defense needs have been brought to lime light by them.

New millennium Indian Technology Leadership Initiative (NMITLI) is under the umbrella of CSIR, which works with private partnership for Nano projects.

Apart from the support rendered by the Central government the role played by the State government has to be stimulated to obtain many successes in Nano. Their fund allocation for local research bodies and institutes like SASTRA in Tanjore, Tamilnadu in South India will encourage many youths to plunge in this field.

Nano Technology Applications

Highly talented Indian defense scientists with meritorious record in war field are selected and deputed even for more than a year in foreign defense research institutes to gain wide knowledge in applications of nano technology for war field attacks.

Nano Technology has gained importance in vehicles, aerospace, logistics, naval vessels, satellites, defense weapons, security, welfare of human beings and sophistication in military applications.

This technology is nothing but knowing and control of matter at dimensions of 1 to 100nm, which enable it to use for sophisticated applications. A nanometer is 10^9 of a meter.

Nano technology comprises imaging, measuring, modeling and manipulating the matter, with this length of scale. In this level, the physical, chemical and biological behavior of material will vary in fundamental and valuable ways from both the properties of individual atoms and molecules or bulk matter. Hence, Nano technology experts are engaged in creating better materials and devices with unique systems that exploit these new properties.

These peculiar properties of Nano technology are obtained through:

1) Small dimensions, which facilities high speed and greater functional density. This is applicable for Nano electronic devices used in defense gadgets.

2) Small, minute and light weight devices suitable for sensors used by army men.

3) High sensitivity phenomenon, which is utilized for manufacturing sensors and nano wires applicable to defense equipments.

4) Special surface effects implemented in army needs, which need lotus effects.

5) Very large surface area offering reinforcement and catalytic effects as demanded by defense experts for their soldiers.

6) Quantum effects like highly efficient optical fluorescent quantum dots suitable for soldiers armaments.

7) Totally different molecular structures, which possess new material properties, Nanotubes, Nano fibers and Nano composites with great strength, which are widely used for army equipments.

Scaling and miniaturization are adopted for creating Nano structures. This is formed as Top down Technology.

Lithographic patterning, embossing or imprint techniques are used. It is later followed by etching and coating strips.

In defense we utilize this technique for devices designed with Micro and Nano electronics, Micro Electro Mechanical Systems, Lotus coatings, Flat Batteries, Solar Cells, Nano Clay Platelets and Tubes.

War field equipments with Nano fibers by electro spinning also have this Top down technology.

Other method to create Nano structure is Bottom up method. By this we construct Nano structures through atom by atom or molecule by molecule. It needs wet chemical or vapor phase processing method like atomic layer deposition. It is also observed that atomic or molecule manipulation is executed by optical, electrical or mechanical nano probes.

Some of its applications are:

-Defense gadgets use carbon nano tubes created by gas phase deposition

- Quantum dots,

- Nano wires produced out of metal oxides and ceramic, polymer type through gas deposition.

This type of Bottom up is created through self assembling molecular and bio structures. Defense wing fully utilize these Nano structures for their warfare.

Though the market growth of Nano tech products volume in 2005 was 100 billion, it has crossed 1500 billion in 2016. Indian Defense Chiefs heartily welcome this trend as they feel that they are safer to face the enemy attacks along the border with military equipments designed with Nano technology. Budget allocation in leading defense institutes like DRDO labs has been enhanced for extensive research in creation of nano structure by Top down and Bottom up technology. Extensive training facilities are arranged with leading universities and labs in Israel, USA, Russia and U.K.

Highly talented Indian defense scientists with meritorious record in war field are selected and deputed even for more than a year in foreign defense research institutes to gain wide knowledge in applications of nano technology for war field attacks.

Production of items in nano level is ideal for mass production at lower costs. It is suitable for Indian defense sector, which has 1,325,000 active warfare frontline personnel and 2,143,000 active reserve personnel.

The gadgets systems and devices made out of nano technology in large scale will be ideal for Indian battalion with 6709 Armored Fighting Vehicles, 290 Self Propelled Guns, 7414 Towered Artillery, 292 Multiple Launch Rocket Systems, 679 Fighters cum Interceptors, 857 Transport Air Crafts, 318 Trainer Aircrafts, 650 Helicopters and 19 Attack Helicopters.

It also caters the needs of Indian Naval wing, which has 2 Aircraft carriers, 14 Frigates, 10 Destroyers, 14 Submarines, 135 Coastal Defense Crafts and 6 Mine warfare weapons.

"Though the defense budget as per Indian rulers is $40,000,000,000, the sophistication of Indian Military equipments needs more liberal allocation" say the defense experts.

Small scale industries are also invited by the Indian government to participate in defense production using Nanotechnology. The defense equipments produced earlier in macro level are now replaced by products done by miniaturization to Nano level.

Institutes for Nano

Indian scientist and institutes show keen interest to have partnership with South Korean Advanced Nano Fabrication Center and Nano Technology Integrated Centers to update knowledge in Carbon Nano Tube applications, Flat panel displays, NEMS devices, Nano electronics and memory, Nano structures etc.

India aims to allocate funds for hybridization, connectivity, assembly, power and micro energy and nano material research works.

Along with government research organizations, the kings of private companies in India like Reliance, Tata and Birla have also joined hands with the Nano Technology developments.

BITS a leading technical institute located in Pilani (Rajasthan), which is under the control of Birla group does an exemplary research studies on Nano Science and Nano materials for defense applications. TERI under the regime of Tata group has experts in Nano Technology to submit research papers in defense conferences conducted within India and in universities located in countries like Japan, USA, U.K. and Korea. Reliance take a major share in producing defense equipments and their team is interested in new design to suit our soldiers.

They prefer tie ups and MOU with companies like Toshiba, Sony, Fujitsu NEC Xerox and Hitachi in Nano technology. These companies have specialized in Nano instruments, Nano electronics, Carbon nano tube technologies ideal for warfare gadgets

As seen in China, where a good platform is formed through Chinese Academy of Sciences to speed up the activities for commercialization of Nano Science and Nano Technology, Indian defense sector also plans with huge budget.

Mass production of Nano structures done in China has to be followed with top priority as we have to face them at any moment along the border.

Indian scientist and institutes show keen interest to have partnership with South Korean Advanced Nano Fabrication Center and Nano Technology Integrated Centers to update knowledge in Carbon Nano Tube applications, Flat panel displays, NEMS devices, Nano electronics and memory, Nano structures etc. The South Korean government has chalked out 10 year Nano technology program which can be adopted by Indian defense sector.

Links with ITRI in Taiwan to promote industrialization of Nano materials will be an added advantage for Indian growth in GDP.

Also tie ups with Academia Sinica is needed for gaining insight in fundamental research in Opto electronics, Non volatile memory and Nano materials.

South Korea holds 4th position among the countries in the world in public funding for Nano.

Taiwan ranks 10th and work far better than Indian experts. So, useful sharing of ideas with these countries to enhance the defense facilities is to be done.

Support from 29 Nano research centers in USA under the control of NSF, DOD and NASA to focus on Nano technology will positively create encouraging trends in Indian defense sector. Routine visits and participation in conferences held there will bring a turnaround in Indian military systems.

These reputed Nano research centers will update our awareness on nano tech risks faced by soldiers, nano scales & systems, nano manufacturing techniques, metrology and standards for nanotechnology. They also focus on fundamental nano scale phenomena and process.

Their guidance for national defense and home land security through control, command and efficient communication is note worthy. Further for surveillance, automation and robotics application for minimizing exposure war fighters, weapons, first responders etc. Further their technical knowhow are to be followed by Indian army men. All their instruments based on nano technology act speedily with accuracy at the same strong and light weight.

Multifunctional materials, small compact planes enable our Air Force mightier than our enemy neighbors. Fully automated self guided UAVs for surveillance are added in Indian military strength only through the support from US experts in Nano.

Thermo electric converters catalysts for excellent conversions, batteries with accurate and unique performance, solar cells and fuel cells used by army wing in India are solely due to tie ups with leading research centers in USA under NSF, DOD and NASA. Our strength in aerospace fully depends upon Nano technology only.

Our border security forces tolling in risky environment use the information technologies adopted by scientists in DNA to enhance computer speed, reduced power consumption and further sealing of Nano electronics. Other advantages are flexible and flat displays, molecular electronics development out of Nano research activities carried out in these institutes in USA.

New Green Technologies adopted to eliminate pollution is another highlight of miniaturized Nano structure developed by experts in this highly developed nation. Remediation and clearing contaminants are other special features seen through their Nano technology.

Innovative steps to create sensors arrays for quick diagnostics of the wounded soldiers is the highlight of nano technology works by USA team members.

Composite structures for replacement of tissues are considered ideal remedy for war field patients. Highly effective medicines to heal wounds, charred organs and bullet injuries are appreciable move by Nano research scholars to assists our warriors.

Facilities to carry the army men with safety and swiftness in any rugged conditions are also provided from Nano Smart Helmet for soldiers at war field through nano is a plus point for safety.

Anti-ballistics materials for these helmets add confidence to face any enemy attacks. Further, the provision of anti-ballistic materials for suits worn by soldiers supports them to march towards opponent' camps and hide outs with caution and protection.

This is possible only through Nano inventions.

Self BC decontaminating nano fiber fabric and health care attention in their suit gives immediate rescue from dangerous attacks it informs the head quarters about the criticality and rush medical team to save their lives. Treatment to wounded battalion is possible with ease. Ventilation and insulation are also done to manage precarious health condition of the affected soldiers.

Sensor tags, Nano fibers and Nano medicines are great contribution from research institutes like IITs in India with foreign aid.

****

Indian Institutes, Scientists and Current Research Projects in Nanotechnology

Many defense related research projects on nano technology under taken by the scientists are highly encouraged by the Government of India and Punjab University

Indian defense sector is very much worried about the welfare of the soldiers toiling in high altitudes.

They are succumbed to lung thrombotic clots and frost bites. Hence nano spray gel for injuries and other remedies are of great concern for nano scientists in India. Further, the application of nano composites to improve the shell life of food items they carry with them in remote areas play a vital role in all defense research works.

Defense Institute of High Altitude Research (DIHAR), an important lab attached to DRDO is located at high altitudes of Leh.

The Border Security Forces are taken care of by this reputed lab experts. They conduct researches to manage the cold desert areas, formulate herbal prophylactics and neutraceuticals to enhance their stamina in hazardous regions. They also think about strategies for good utilization of non-conventional energy sources.

DIHAR also has a lab at Chandigarh to conduct study in the topics meant for safeguarding our soldiers.

Apart from this institute, another leading institute is Institute of Nano Science and Technology (INST), is at Mohali. This is a leading institute of DST, mainly aiming to conduct research projects on environment, bio-sensors with nano technology and nano science.

A MOU has been signed between these two giant institutes of nano technology to study further on following subjects:

1) Formation of Joint Technology Development Council
2) Sharing research facilities
3) Nano packing materials for soldiers especially in remote areas

4) Knowledge-sharing with leading university experts and foreign scientists involved in defense

Punjab University has also joined hands with INST to go in depth in the field of nano technology. 14 premier research institutions in Chandigarh are involved in innovation and encouragement to research scientists in India. They are IIT Ropar, CSIO, GMCH, BPU, NABI, TBRL, PGIMER, ISB, PEC, NITTR, IISER, IMTECH, NIPER and INST.

They share their knowledge on nano development in India under the regime of Chandigarh Region Innovation and Knowledge Cluster. Many leading scientists play a vital role in this great venture.

They are as follows:

Professor Ashok. K. Ganguli, P, IISc, Bangalore Director, email:director@inst.ac.in, instmohali@gmail.com.

Abir de Sankar, Associate Professor, Scientist-E, email:abir@inst.ac.in

Asifkhan Shanavas, PhD, IIT, Mumbai, Scientist B, email:asifkhan@inst.ac.in

Ashish Pal, Ph D. IISC Bangalore, Associate Professor (Scientist E) email: apal@inst.ac.in;

Bhanu Prakash, MSc, IIT-Kanpur, Scientist-B, Email:bhanup@inst.ac.in;

Chandan Bera, PhD, École Centrale Paris, France, Scientist B, Email:deepika@inst.ac.in;

Ehesan Ali, PhD, IIT Mumbai, Associate Professor (Scientist E), email: ehesan.ali@inst.ac.in;

Jayamurugan Govindasamy, PhD, IISc, Bangalore, Asst Professor (Scientist-D) email:

jayamurugan@inst.ac.in; jayamurugan.inst@gmail.com;

Jiban Jyoti Panda, Ph D, LCGEB, New Delhi, Scientist C, Email: jyoti@inst.ac.in ;

Kamala Kannan Kailasam, Ph .D, Univ of Stuttgart, Germany, Associate Professor Scientist E, Email id: kamal@inst.ac.in;

Kaushik Gosh, PhD, Meijo University, Japan, Asst Professor, Scientist D, Email: kaushik@ inst .ac.in;

Kiran Shanker Hazra, PhD, IIT, Bombay Scientist C, Email ID: kiran@inst.ac.in ;

Manish Singh, PhD, University of Allahabad, Scientist B, Email ID: manish@inst.ac.in ;

Menaka Jha, PhD, IIT Delhi, Scientist B Email ID: menaka@inst.ac.in ;

Monika Singh, PhD, IIT Delhi, Scientist B, Email ID: monika@inst.ac.in ;

Mukesh Raja, MSc, JMI, Delhi M.B.A; Sikkim Manipal University, Scientist B, Email ID: mukeshraja@inst.ac.in ;

Neha Sardana, PhD, IMPRS-MLU, Germany, Scientist B, Email ID:nsardana@inst.ac.in;

Prakash. P.Neelakandan, Ph. D, CSIR-NIIST, Thiruvananthapuram, Associate Professor, Scientist E, Email id: ppn@inst.ac.in;

Priyanka, PhD, MTech, Chandigarh, Scientist C, Email ID: priyanka@inst.ac.in

P.S.Vijayakumar, PhD, Agricultural University, TN, Scientist C, Email ID: psvijayakumar@inst.ac.in ;

Rahul K. Verma, PhD, CDRI, Lucknow Scientist C, Email ID: rahulverma@inst.ac.in ;

Ramendra Sundar Dey, PhD., IIT Kharagpur, Scientist B Email ID: rsdey@inst.ac.in;

rehankhan@inst.ac.in ;

Sangita Roy, PhD, IACS, Kolkata, Scientist C, Email ID: sangita@inst.ac.in ;

Sanyasinaidu Boddu, PhD, BARC, Mumbai Scientist B, Email: sanyasinaidu@inst.ac.in;

Sharmistha Sinha, PhD, IISc Bangalore Assistant Professor (Scientist D) Email ID: sinha.inst@gmail.com;

Shyam Lal M, PhD, BHU, Varanasi Scientist B, Email ID shyamlal@inst.ac.in;

Sonalika Vaidya, PhD, IIT Delhi, Scientist C, Email ID:svaidya@inst.ac.in;

Subhasree Roy Choudhury, PhD, CU, Scientist B, Email ID: subhasreerc@inst.ac.in;

Sucheta De, PhD, CGCRI, Kolkata, DST Young Scientist, Email ID:suchetade@inst.ac.in

Surajit Karmakar Ph. D: University of Calcutta, Associate Professor Scientist E, Email:surajit@inst.in;

Suvankar Chakraverty, PhD, S.N. Bose National Centre for Basic Sciences, Assistant Professor, Scientist D, Email ID: Suvankar. chakraverty@inst.ac.in;

Tapasi Sen, PhD, IACS, Kolkata Scientist, C Email ID: tapasi@inst.ac.in;

Vivek Bagchi, PhD, IIT Delhi Scientist B, Email ID: bagchiv@inst.ac.in ;

Professor Ashok k Ganguli (FASC, FNASC and FRSC) Director, Email ID: director@inst.ac.in, instmohali@gmail.com, Phone No. +91-172-2210073

Shri Umesh Chandra Prasad, Chief Finance and Administrative Officer, Email ID: cfao@inst.ac.in , phone No. +91-172-2210074

Shri Praveen Kumar Datta, Consultant (Project Management / Dean of students / CPIO, RTI) (Formerly Director, Department of Electronics and Information Technology (DEITY), New Delhi), Email ID: pkdatta@inst.ac.in , pkdatta1950@gmail.com admin@inst.ac.in, Phone No. +91-172-2211074.

Shri Ashok Kumar Kakria, Consultant (Finance and Accounts); Store & Purchase Officer (formerly Chief Administrative cum Finance Officer, Centre for Development of Advanced Computing, Mohali),

Email ID: kakria@inst.ac.in, kakria@gmail.com, Phone No. +91-172-2210075

Dr.Surajit Karmarkar, Dean of Academic Affairs. Email: surajit@inst.ac.in, Phone no: +91-172-2210075

Dr.Sharmistha Sinha, Dean of Research Email ID: sinhas@inst.ac.in, Phone .No +91-172-2210075

Dr.Sangita Roy, Associate Dean of Students, Email ID: sangita@inst.in, Phone no. +91-172-2210075

Dr.Sonalika Vaidya, Associate Dean of Academic Affairs, Email ID: svaidya@inst.ac.in, Phone no. +91-172-2210075

Dr. Suvankar Chakravarty, Associate Dean of Research, Email ID: suvankar.chakraverty @inst.ac.in, Phone no. +91-172-2210075

sumansharma76@gmail.com, Phone No: +91-172-2210075.

Shri Surindar Singh, Senior Supervisor/ Caretaker (Formerly with Indian Air Force) Email ID: surendersingh@inst.ac.in, phone no. +91-172-2210075.

Shri Dhanjit Singh, Office Assistant, Email ID: dhanjit@inst.ac.in , Phone No: +91-172-2210075.

Many research projects on nano technology under taken by scientists are highly encouraged by the Govt of India and Punjab University.

Some of them are:

Ms.Rashmika Singh - Neurotoxicity of Nan particles

Ms.Manisha - Oncolysis by Targeted Therapy with Trojan horse bacteria.

Ms. Pranjali Yadav - Graphite Carbon Nitride Nanostructures for biomedical applications

Ms. Ankush Garg - Exploring the cellular determinants of p53 aggregation in

Cancer

Ms. Geetha - High frequency ultra –thin Graphene resonator based mass detectors

Ms. Harsimran kaur - Supramolecular Hydrogels and Tissue Engineering

Ms. Prabhjot kaur - Electronic and thermal properties of strong spin-orbit coupling systems

Ms. Anas Ahmad - Synthetic Lethality and Centre- Targeted Centre Therapy by Small Molecule Inhibitors / Nano-Formulations

Mr. Jojo. P. Joseph - Synthesis and applications of polymeric nanoparticles

Mr. Pushpendra - Photo catalytic water splitting by semiconductor nanomaterials

Ms. Taru Dube - Development of biocompatible nanostructures capable of traversing the blood brain for targeting glioblastoma/ braintumors.

Mr. Pulkit - Mesoporous Silica Nan particles for controlled and targeted nutrient delivery in plants.

Mr. Nityasagar Jena – Computational studies on nano materials for gas sensors.

Mr. Anirban Kundu - Opto-electronic properties of Graphene

Mr. Neha Wadhera - 2DEG at the interface of a strong spin orbit coupling perovskite Oxide.

Mr. Atul Dev - Synthesis and Characterization of Nan composites and their usage in Nanotherapeutics.

Ms. Renu Rani - Development of MoS2 based Opto-electronic devices and sensors.

Ms. Harmanjit Kaur - Bio-receptor functionalized nanostructure sensing platform.

Mr. Sandeep - Eccentric inorganic polymeric nanoparticles for plant disease control in complementary fashion.

Ms. Ruchi Tomer km – Electronic and magnetic properties of oxide interface and super lattices.

Mr. Soumen Ash - Thermal properties in nanostructure materials.

Ms. Rashmi Jain - Bio-inspired Hydrogels for healthcare

Ms. Ritu Rai - Mixed carbides nanostructures for electro-catalysis

Mr. Ashmeet Singh - Smart Gels

Mr. Ankur Sharma - Pulmonary Delivery of Antimicrobial peptides using porous Nano particle aggregates (PNAPs) against Pulmonary Tuberculosis

Mr. Dimple - Computational modeling in energy relevant nano-materials and Molecular electronics

Mr. Naimat Kalim Bari - Cell free Bio-reactors from the shell proteins of bacterial micro compartments

Ms. Swati Tanwer - Plasmonics nanostructures based on DNA origami DNA-directed self-assembled nano-antennas to get strong Fluorescence Enhancement for Bio-molecular Assays and Sensing Applications

Mr.Munish Shorie - Low cost immunodiagnostics for cardiac management

Mr. Rajender Kumar - Nano-structured materials for electro and photo electro chemical applications

Dr. Madhunika Agarwal - Protein based nanoparticles for Drug Delivery in cancer

Dr. Vinod Kumar - Applications of nano-materials for theranostic purposes

Dr. Gaganpreet - Dynamical behavior of fluids inside nano-channels

Dr. Manu Sharma - Metal Oxide Nano particles for Catalytic Applications

Dr. Seema Gautam - Atomic scale investigation of activation and adsorption of hydrocarbon over metal clusters through first principles electronic structure method.

The efforts taken by them will obviously boost confidence in the minds of Indian defense sector that there are scientists and institutes to take initiatives in the field of nano technology and serve the nation in the long run.

Functioning of Nanotechnology in India

Nano Science and Technology Initiative and Nano Science Technology Mission do well in innovative research works in defense sectors to support the suffering soldiers in India.

There are 3 major bodies for Nano technology developmental works in India for military force. They are Nano Technology Generation bodies, Knowledge Transfer bodies and Knowledge Application bodies.

Under Nano Technology Generation unit we have R&D divisions to monitor and support all universities and colleges in India.

Research institutes controlled by both public and private are shouldered to update in Nano technology in defense applications. Then we have centers of excellence like Unit of Nano Science, Center for Nano Technology, Centre for Computational Activities and Centre for Material Science.

TERI, NISTADS etc, which cover social science, framing of policy and research works are under this knowledge generation bodies. Environment Health and Safety (EHS) and research works of National Institute of Pharmaceuticals and Research (NIPER) are also a branch of this body.

Knowledge transfer bodies have Centre for Technology Transfer and NT-NCL in Pune to cover specialized Incubators and S & T Entrepreneur Parks.

Knowledge Application bodies have Confederation of Indian Industry, Federation of Indian Chambers and Industry (FICCI) and Associated Chamber of Commerce (ASSOCHAM), the oldest, leading and largest apex body of Chambers and Commerce to support them. They conduct international symposiums, conferences and discussions to enable international talented scientists and entrepreneurs to interact with Indian counterparts and businessmen involved in Nano Technology.

The members of these bodies are engaged in product development.

This scenario of Nano science and technology in India is supported by media (both print and visual).

Policy makers, Regulatory bodies and Authorities play a vital role with Department of Science and Technology (DST), Department of Bio Technology (DBT) and Council of Scientific and Research (CSIR) members. They are governed by Ministry of Science and Technology (MoST).

We have Ministry of Communication and Information Technology (MoCIT) controlled by Department of Information Technology. Ministry of Health and Family Welfare under Indian Council of Medical Research (ICMR) guide the ex-army men family members and existing battalions.

DRDO in Ministry of Defense Ministry of Agriculture , Indian Council of Agricultural Research (ICAR), Ministry of New and Renewable Energy, Department of Atomic Energy, Ministry of Commerce & Industry, Ministry of Water Resources (MoWR), Ministry of Food Processing Industry (MoFPI), Ministry of Environment and Forests (MoEF) and Ministry of Textiles are under policy makers unit.

Civil Society and Community Organizations, is another branch to develop Nano Science and Technology in India. They are engaged in Gene Campaign for influencing policy and create awareness among the society.

NIMBKAR Agricultural Institute in Pune executes technical dissemination activities and advices the wounded ex-army men for cultivation and further survival.

India Nano acts as a bridge between academia and industry. It is supported by liberal funds by Indian companies.

Financial sector also act to develop Nano Science and technology in India through venture capital, public funds, private funds and international funds.

Project sanctioned for development of Nano technology was initially 22 in the year 2002-2003 and it gradually rose to 35 in the year 2003-2004.

During the tenure of Dr.Abdul Kalam, as President of India, many scientists were encouraged to participate in international conferences and useful tie-ups were witnessed in this field. More than 50 projects were recommended and executed successfully under his leadership as he was a leading scientist.

It was a boon period for defense experts, stalwarts and Nano researchers.

Nano Science and Technology Initiative and Nano Science Technology Mission do well in innovative research works in defense sectors to support the suffering soldiers in India.

Centers of Excellence for Nano Science and Technology under the care of Nano Science and Technology Initiative concentrate on defense application projects and finish the work in a stipulated time to equip the soldiers with the desired gadgets.

****

Future Nano Products

Nano Technology Competence Centers will be established in leading cities like Delhi, Mumbai, Kolkata, Chennai and Gujarat to promote Ultra Precision Surface Engineering, Nano Coatings, Nano Optics, Nano Bio-technology and Nano Chemistry to support the Army, Naval and Air Force sectors.

There are still many unknown products, which can be produced out of Nano technology, even though many research labs have toiled to produce new nanostructures and applications.

At present the Defense research labs have joined hands with institutes like IIT (Delhi, Kharagpur, Mumbai and Chennai) ISRO, SRM University, SASTRA University and Anna University to produce Opto electronics, CCD Chips, Processor Chips and gadgets based on Nano electronics for army men. They have also produced MEMS devices, Inertial Motion Units, Pressure and Flow Sensors.

Along with these they have contributed nanoclay reinforced materials, nano sized pigments and additives to act as UV Blockers in paints, lubricants, coatings etc.

Fuel cell components, Membranes and DNA Biochips are also listed as contributions from Nano technology experts from leading research labs in India.

As the Indian government is keen to send our Defense Chiefs and their subordinates to various Universities in U.K. USA and Israel, much is expected in Nanotechnology fields to assist the Indian soldiers. Their contributions will be seen in producing μ poser to suit war field attacks in remote areas. Distributed, Wireless and Autonomous Sensors will also pave way for better handling of enemy missiles.

Adaptive materials with built-in-sensors and actuators will enlighten our defense scientists to develop gadgets with fast response and utility.

Bio transistors is another development expected, which will support defense instruments with more sensitivity.

Nano wire and Nano bio will also play a vital role in future developments

Nano medicine will enable new treatment to wounded soldiers, who need emergency treatment and quick recovery from bullet wounds.

Technology Radars covers the applications like bio & life science, energy & power, information and communication, nano materials and manufacturing sector etc.

Now Nano technology radars are considered for defense in the field of vehicles, space satellites, weapons, security for battle field and soldiers, naval force, medical support for army men, operating fighters & aero planes and logistics.

Indian Defense experts are planning to establish micro satellite, micro factory, micro invasive devices, micro life, distributed sensors and actuator systems to be used in modeling, designing, micro systems and for many hazardous situations.

Much fund and support from DRDO is needed to produce wireless autonomous sensors. They also involve experts to analyze flexible electronics in a foil for lighting, tags, sensors, etc.

Indian government should arrange National Nano Technology Programs as done in European countries to encourage defense scientists to update their knowledge.

Action plans have to be chalked out for future defense innovative research works on Nano instrumentation, nano manufacturing technologies, regenerative medicines for soldiers at war front and nano particles technology.

Special concentration on risks on health due to usage of Nano products by warriors in hazardous areas is a must.

Nano Technology Competence Centers will be established in leading cities like Delhi, Mumbai, Kolkata, Chennai and Gujarat.

Indian defense experts are keen to promote Ultra Precision Surface Engineering, Nano Coatings, Nano Optics, Nano Biotechnology and Nano Chemistry to support the Army, Naval and Air force sectors. A good network is also planned allover India to create awareness among young scientists about Nano technology.

At least 2500 experts are to be created for future plans to take shape in University of Rajasthan, Alagappa University, Amity Institute of Technology and Amrita Centre for Nano Sciences, VIT University, Jamia Millia Islamia Centre for Nano Science and Nano Technology, Jawaharlal Nehru Technological University, Noorul Islam University and National Institute of Technology.

More numbers of talented Muslims are from Jamia Millia Islamia Centre and Noorul Islam University. They are well versed in Urdu and Pakistani local languages. Their training activities in Iran and Arabian countries with Nano experts have fetched fruitful results. All the above Universities have links with DRDO and Defense Chiefs, so that the desired research works are done to satisfy the needs of Indian warriors.

The Defense advisors clearly explain the difficulties faced by them in the warfront and suitably design the equipments with Nano Technology. Group discussions and critical analysis of the pit falls faced by soldiers are conducted to get the accurate results. Their aim is to see that the end users are comfortable at the same time more sophisticated than the neighboring enemies like, Russia, China and Pakistan.

They have also taken utmost care to see that their investigations and research works are not leaked out to media or any terrorist spy groups. So, well educated and at the same time reliable military men are deputed for discussions and implementations.

Micro systems and assembly methods followed are kept confidential.

Required facilities are arranged in all the above Indian Universities with strong security checks.

****

International Co-operation for Nano Technology and Nano Composites in India

Especially for military applications, many more research works are still pending and the Indian experts need financial and technical knowhow.

Development of Nano technology and Nano Composites in Indian scenario is impossible without the support of reputed foreign scientists, institutes, companies and R & D Centers.

Especially for military applications many more research works are still pending and the Indian experts need financial and technical knowhow.

This new Nano technology will open up new avenues of research to enhance the competitiveness of our Indian industries and also create new products that will pave way to positive changes in the lives of our citizens, who are economically backward,. This will boost various applications unknown so far.

Indian companies, which strive to produce defense equipments, sensors, radars etc with international support, are;

Ashok Leyland, Bharat Electronics (BEL), Hindustan Aeronautics Ltd (HAL), Mazagon Dock Ltd., Tata Advanced Systems with Boeing, Sirkorsky Aircraft Corporation & Israel Aerospace Industries, Electronics Corporation of India Ltd., Spectrum Info Tech, Bharat Earth Movers Ltd., Bharat Dynamics, Rotta, Mishra Dhatu Nizam Ltd. Auro Integrated Systems, Bharat Forge, Goa Shipyard, Mahindra Defense Systems, Bae Systems & Sea bird Aviation.

Adnano Technologies Pvt. Ltd, located in 307, 3rd floor, KEONIC IT Park, Machenahalli Industrial Area, Shimoga, 577222, Karnataka, produce Graphene and Nanotubes, which play a significant role in defense equipments, weapons and sensors. They also serve as consultants for Nano materials. Apart from producing Multi-walled Carbon

Nanotubes, they extend any critical services for AFM, XPS, Contact Angle, Zeta Sizer, TEM, FTIR, XRD etc.,

Icon Analytical Company P Ltd, located in 40th cross road, 5th block, Jayanagar, Bangalore 560041, Phone: 08032958767, imports analytical instruments based on Nano Technology from USA, Japan and Europe. They have their regional offices in Mumbai, Delhi and Kolkata.

Kerala Minerals & Metals ltd manufactures Titanium dioxide nano particles of various grades. Contact address: NH66, Chavara, Kollam, 691001, Phone: 04712687117.They have established Titanium sponge plant, shouldered by Defense Metallurgical Research Lab (DMRL) and Vikram Sarabhai Space Centre.

Titanium oxide they produce is used for Indian Fighter Aircraft due to its light weight. They also cater to Indian nuclear plants and industries involved in heat exchangers and reactors. Rutile, Silliminite, Zircon etc are some of their product widely marketed to Asian countries and USA. This plant was inaugurated by Defense Minister in 2011, as the Indian defense division is their major customer. More than 2200 are employed in this organization, managed by Kerala State Government.

Polymerized toners are manufactured by Navran Advanced Nano Products Development International P Ltd, B 10, GF, Park Centra Building, Sector 30; Off Exit 8 on NH8 Opp to 32nd milestone, Gurgaon, Haryana, 122001.This is a subsidiary of Nano Product Development Inc USA. They have shouldered with Oxonica to market Envirox, a diesel additive to improve fuel efficiency and eliminate green house gases. Their plant is in Dhamandri District, Una, Himachal Pradesh, with manpower of 55.

High quality smart polymers Nano composites and Nano materials supplied by Quantum Materials Corporation P Ltd are used for lighting, telecommunication, energy industries and electronic gadgets engaged by Indian military warriors. Their contact address: #190, 2nd floor, 9th cross, HMT Layout, R.T. R.T Nagar, Bangalore 560032, Phone: 91 080 40914091.

Auto Fiber Craft Powders P Ltd is a company engaged in producing nano materials. Their Nano size silver powder is used in electronic applications like conductive inks and pastes etc. Their product is RoHS compliant and internationally appreciated. Its production unit is located in Industrial Area, Gamharia, Jamshedpur, Jharkhand, and Phone: 09955394798. They are manufacturers of FRP Components, Rotomolded products, Nano silver powder Nano materials and Nano Gold powder for the past 25 years. They focus on automotive and mass transportation industries with vacuum bags, SMC compression molding, VARTM, RTM etc.

Their technical scientist staffs are from Indian Institute of Technology, Kharagpur, Delhi, Mumbai and Kolkata. Contact office address: Auto Fiber craft, C-17, Phase-6, Adityapur 2nd Area, P.O Gamharia, Jamshedpur, 832108, Phone: +91-9234621160. Email: autofibercraft@gmail.com. Their Nano Palladium, Nano Ferrite and Nano silver are marketed to countries all over the world like USA, Japan, UK, S.Korea etc, as they possess quality. Total man power is 50, say their representatives.

Carbon Fiber reinforce plastic filament wound tubes (ID=6, L=2m),FRP window frame, concealing fittings, FRP covers, Panels for sidewall and end wall FRP air inlet ducts, FRP mudguards, Engine hoods, Glove box lids, Fenders mud flaps, FRP Cooling towers and FRP Water tanks are their products marketed for Indian and overseas clients.

Superfine Nano Silver powder 30-54nm ≥ 99.9%, Ag, Silver powder 99.9% Ag, Silver powder, purity ≥99.9% AG, -200 mesh Silver Pulver, Reinheit 99.9% Ag,-200 mesh Pure Silver Powder - Flake Morphology (APC1 micron) and Nano silver coated copper powders for conductive rubber applications etc they produce have ISO 9001-2000 certificates.

Many varieties of Nano silica products are manufactured by Bee Chems. They have specialized in silica and alumina industries.

Bee Chems is located in E5, Panki Industrial Area, Site-1, Kanpur, Uttarpradesh, 208022, Phone: 080-21345678.

This is the first company to produce Colloidal silica, activated desiccants and eco friendly silica gel. They market in India as well as in USA, UK and all Asian countries. Their desiccants for moisture protection controls are used for defense sector. Also fine silica produced by them is used for clinical applications. They are the producer of maximum types of desiccants in the world.

Sorba Pouch, Pura Fresh-Food fresh products, Protect-it-bags for moisture sensitive powders, Ethylene absorbers and protection products, Dew pouch-activated desiccant packs, Silica Gel Pouches, Nano Silica Binders etc are exported to their customers in UK, USA and Europe. They produce alumina modified sols, cationic sols, modified nano sols and pure alumina sols. For exports details contact No. 9919662259.

Graphene and Carbon Nanotubes are manufactured by Bottom up Technology Corporation.

This reputed company also supply nano materials like Graphene & MWCN.They render rapid and integrated solutions for nano composites, defense, material science and energy sectors.

Graphene they supply has:

Purity ≥ 96-99%

Thickness – 3-6 nm

Surface area – 323-6co m2/g

Diameter – 10-20 micron

Varieties of Graphene they supply are: Dispersion Graphene, OH, Functionalized Graphene, NH2 functionalized Graphene, COOH Functionalized Graphene and Catalyst for CVD Graphene etc.

It has its registered office in M Pipra, MBDih, Mahagama, Godda, Jharkhand, 814154. They have their branches in Bangalore, New Delhi, Mumbai and Ghaziabad.

They have clients all over the world with an experienced team of marketing experts. Contact email btc@bt-corp.com.

NanoShell produces 350 types of Nano powder. Their Carbon Nanotubes metal and Alloy Nano powder and Oxides Allied Nano powder have wide global market network.

50 types of Nano materials are produced by Nanoshel. Further MWCNT, Nano tubes and SWCNT are their major products.

These products are marketed by Intelligent Materials P.ltd, Village Sundran, Mubarakpur-Sundran Road, near Parabolic Drug Dera Sassi, Punjab. Pin 140507.

They have trade links in Taiwan, U.K. Thailand, China, Nepal, Iran, Japan, Saudi Arabia, Kuwait, Malaysia, Iran, Singapore, Europe and USA.

They market Quantum Dots, Micro powder, Carbon Nanotubes, Nano Powder, Nano Fabric, Nan Wires, Nano Rods, Doped Nano powders, Compound Nano powder etc.

Nilima Nano Technologies are involved in producing and marketing Nano technology based coating with protective nature needed for defense equipments.

This reputed company is located in 27, AR Rangnekar Road, Gamdevi, Mumbai, 400007. Contact: +91-09920987611.

Their nano coating has self organizing ceramic nanostructures with the size of 10-50nm. When we coat this, we get a seal on the surface to reject water, dirt or oil. They are highly resistant to acids and cleaning products. Hence they are ideal for defense weapons, equipments and applications.

Their coating is UV-light stable and coated area is resistant to pressure up to 60bar and temperature up to 500°c.

Indian Defense sector prefer their coatings.

Integrated Product Development in India

The Indian Defense Chiefs in Air force, Navy and Army expect their men to possess ability to meticulously integrate technical performance, cost and schedule needs into a single job pack.

Integrated Product Development is a unique way of producing equipments and components to meet all technical and performance requirements with the desired cost and time schedule.

Indian Government is keen in implementing it.

They take action to even eliminate unworthy staff with voluntary retirement scheme. Modi said "I will not sleep and also allow others to sleep". This is 100% true in defense sector. Everybody is expected to have data in their finger tips to face enemy attack at any juncture.

The Defense Chiefs in Air force, Navy and Army expect their men to possess ability to meticulously integrate technical performance, cost and schedule needs into a single job pack.

It should be important to see that it is done with complete traceability to customer requirements throughout the life cycle.

The soldier at the war front should be satisfied for the entire life cycle.

Low weight, easy handing and quick replacement of spare parts without complication are the aims of Indian defense production team.

The defense team should have the capabilities to plan and execute them with techniques like Configuration Management of designs and documents.

It is pertinent that Program Management activities are carried out with the guidance of senior or retired Army Generals and Navy experts.

Virtual verification of technical details, production schedules and in-service performance before committing to hardware are essentially done.

Ministry of Defense repeatedly proclaims that our army scientists and entrepreneurs should adopt best practices and lessons and utilize it to create outstanding products and by-products. This will obviously strengthen Weapon Production Management in India. Small Scale Industries involved in it have to face the burden of financial crisis and their products are not up to the mark as expected by the soldiers in the battlefield.

It is essential that we have single master source of all defense data and 3D images that can be utilized for the present complicated aerospace platforms. With this we have to synchronize a global virtual network of designers, developers, production experts, manufacturing technical team, marketing and service groups.

All defense and aerospace organization have a vision for program execution excellence on every program.

Ayyappan Ramamurthy Technical Director of Siemens said "PLM software will satisfy the Ministry of Defense, Government of India as a single easy tool with traceability from program statement of work to schedule to bid to actual performance".

The DRDO and ADE Scientist expect high design change rate. After the radar or missile or UAV is tested, they may not get the anticipated results or parameters. So, they may demand the manufacturer to alter the design values to suit Indian climate or capabilities of our soldiers. Ramamurthy from Siemens said that their software will reduce any late changes as it will verify the technical performance then and there to enable to pass all tests conducted by battle field Indian experts.

Fund allocation for defense in India is low compared to other countries. Hence good supplier relationship management is needed to render to integrate technical, cost and schedule needs.

Indian Aerospace industry is aiming sustainability to reach new scales with life cycle that span 100 years and performance metrics that drive continuous developments in availability, maintainability, overhaul cycle reduction and reliability. The Indian army wing expects continuous configuration management and real time feed back to achieve engineering improvements and betterment of design effectively and speedily. This will enable us to synchronize the supply chain such that spare parts are readily available at the right time and right place.

We need a solution to integrate Indian Defense Project Management, Systems Engineering, Requirements Management, Configuration and Change Management with digital simulation to optimize concept development.

Knowledge Management for the Indian aerospace engineers is essential to get the required training, certification and specialization. Knowledge integration is also a must to support a single and secure environment that offers all program participants with entitled access to program or product information.

Knowledge Visualization, Seamless Integrated Program Planning and Execution Environment are needed for our army scientists and technocrats.

"Integrated Product Development is weak in Indian industries and hence zero error certification is not achievable by them for their products. Due to this trend, we have to import most of the parts from Switzerland, where they adapt good Integrated Product Development Technique. Collaborative business environment is also essential in India.

We need a system to create information once and maintain it its entire source, while enabling all users to access and utilize the data throughout its Life Cycle" says Prasad, Project Director for Rustom-II.

We also need systems, which will enable timely and accurate configuration control, said he.

Our manufacturers should be aware of:

-Import and export regulations

-Security processes and procedures

-Environmental compliance

We should also have a good method for preparing Contract Data Requirements List and Sub Contractor Data Requirements List.

The Indian defense advisors stress that we should adhere to government and data standards for smooth sailing in Integrated Product Development Management in India.

****

Indian Companies Involved in Nano Products

Indian companies involved in Nano seek collaboration with US, UK and Europe based multi-national companies.

At Nano scale, the materials behave in a totally different manner and also much better than the same material or substance at macro size.

Carbon may conduct heat and electricity in a better manner and silver will also have enhanced anti-microbial properties.

Substances at Nano level have other characteristics like increase in capabilities to cope up with higher temperature and pressure increase in elasticity, becoming stronger or weaker, having better chemical reactions etc. This has lead to production of many varieties of products with Nano technology useful for defense.

Leading companies in India, which offer valuable Nano products are:

1) Nano Shel India

2) Yashnanotech

3) Micro Materials P Ltd

4) Nano Bio Chemicals

5) Velbionanotech

Nano Shel India

This has the credit of being the first company to produce Carbon Nano tubes in India.

Their single wall, double wall nano tubes are popular in Indian market. With 70 employees it functions as a unit of Intelligent Materials P Ltd. in Plot 211, Sector 12,Panchkula,Punjab 134112,They have their another contact address: Village Sundran near Parabolic Drugs Dera Bassi, Punjab.

They are involved in Nano Powders, Compound Nano Powder, Nano Wires, Nano Rods, Nano Fabric Coatings, Alloy Nano Powders, Metal Nano Powder, Graphene Nano Powder, Doped Nano Powder, Oxide Nano Powder and Clay Nano Powder etc, to cater the needs of both domestic defense needs and international markets in USA, UK, Japan, Korea and Europe etc. Contact email id: contact@nanoshel.com. Cell: 0977988077. CdSe/ZnS Quantum dots, Inp/ZnS Quantum Dots, ZnSe/ZnS Quantum dots etc are also produced by them. Lithium lon Battery Material, ITO Coated Glass, Carbon Nano Tubes Array etc are also marketed by them.

Yash Nanotech Ltd

Yash Nanotech Ltd is located in Unit 295, Leela Palace, 14th Cross Road, RMS II Stage, DOI, Bangalore, Karnataka, 560094, India. Its Mumbai office is located in 7/10Botawala building, 3rdFloor, 09 Horniman Circle, Fort, Mumbai - 40001.

E mail anuragupta500@gmail.com

This is a branch of Yash Management & Satellite Ltd and they have tie-ups with Cientifica Ltd located in U.K.

With a paid up capital of Rs1, 325, 000, it is managed for the past 9 years by other Directors like Satish Kumar Gupta and Sandeep Kumar Mangal. Contact: +91(80)4217239.

Velbio Nano Tech

With J. Asantraj as its CEO, Velbio Nano Tech functions for the past 14 years in R&D with VBN analyze Genes and Proteins for drug discovery, Nano Chip design and Microbial Nano Chip design. Their core areas are Neuro, Cardiac and Urology. Their drugs are used to cure AIDS, Cancer, Cardiac issues, Kidney Stones etc. These drugs are assembled in Nano Chips to enable the delivering activities in our human body. This phenomenon is widely applied for army men, who are severely wounded in the war front.

Their CEO has wide experience for more than a decade in Information Technology and Pharmaceutical industry like Nano Soft Microsystems, Velnetwork etc. He is assisted by Alice, who strives to reach the goal in global market. Dr. Ajay Kumar Gupta is engaged as Senior Nanotechnology Scientist in this reputed Nano organization. He has rich experience in crusade Laboratories in U.K, as the leading scientist to prepare and characterize Polymer, Magnetic and Inorganic, Ceramic Nano particles to produce new formulations for small molecules, proteins and gene based drugs. As a doctorate from Glasgow and Fellow he has also extensively studied the reaction of his drugs with cells.

Dr. Balaji Chatterjee is another doctorate from Boston University (USA), who is designing Bio Nano Chips with 15years of experience.

Scientific Advisor for this great Nano organization is Dr. Chandrasekhar S. Ratkal, who has a decade of experience in Kidney stones. Many soldiers and ex-army men are attended for stone problems with his guidance.

Micro Materials P Ltd

Micro Materials P Ltd. is engaged in developing applications based on Hydrogen production, Lead Acid Batteries, Automotive Catalysis, Organic Catalytic applications, Replaceable Catalyst Cartridges etc.

Air pollution is a major issue in India. Their catalysts render a good solution to meet the Nox emission regulations.

Reduction of water-gas shift reaction to produce low cost hydrogen is possible with their catalysis. It is a Prototype Wood Chip Based Gasifier to give good quality hydrogen at low cost.

The representative of this great organization says that the quantity of noble metals utilized for their catalysts is only 1/5 of the quantity engaged for identical applications in traditional catalysts.

Their noble metal ion doped Ceria Catalyst powder has organic catalytic reactions. Platinum doped Ceria Catalysts, which is utilizing their proprietary process, are considered as good oxygen hydrogen recombination catalysts. Their office is located in#7,1st Main, 1st Floor, Kodava   Samaj,   Vasanth   Nagar,   Bangaluru560052   , phone:+91-0802289290.

Nano Bio Chemicals

Nano Bio Chemicals located in 1004, Shiv Tapi, HG Road, Gamdevi, 022-23676377, Mumbai has 50 employees and turnover of Rs. 15 crores annually. They are engaged in Pharmaceuticals and Bio-tech research works pertaining to defense needs.

Indian companies involved in Nano seek collaboration with US, UK and Europe based multi-national companies.

"More tie-ups develop Indian GDP growth will also develop" say the Nano scientists'

Nano Technology for Indian Defense

We need more information on the nature and characteristics of applications of Nano materials. Suitable tests are to be conducted by R & D wing in Defense sector to save our soldiers from harmful health havocs. This needs the support of foreign experts in USA, UK and Israel with liberal fund allocation.

To manage this new technology

India aims for developmental research works in Nanotechnology with foreign support.

For that we need a unique infrastructure to encourage R & D and commercialization of Nano products.

There may be many constraints and concerns among the end users. These aspects have to be viewed seriously.

The top leaders and Indian scientists should chalk out the required strategies, policies and strong institutions to manage this new technology.

Strong human resources with skill and training are a must with multidisciplinary perspectives to march towards progress.

Risks

The risks in society faced due to implementation of this new technology had to be sorted out. Awareness programs have to be planned or else huge investment with foreign companies in Nano will go to a waste. Especially rural folks will desist from welcoming new and unknown technology based home products.

Regulatory Boards headed by reputed scientists are needed in India to have a healthy tie-ups in Nano technology based MNCs.

Necessary follow-ups and preparedness are also essential in Indian defense sector. There should not be any dubious statements or actions while implementing this new science for defense. Transparency at all levels is needed.

Academic support

Public involvement should be strictly followed to avoid controversies in the Parliament by opposition parties. Scams should be avoided in money transactions.

Academic support is also vital. Institutions whether they are big or small should spare a division and staff for educating students about Nano technology.

Totally different R&D strategies along with re-orientation of science and technology projects in universities, research labs, funding agencies and industrial sector are needed. They should also have adequate facilities for higher learning and intellectual interactions to promote this fine technology.

Governance bodies

Adaptive and responsive governance bodies have to play a vital role for regulatory application to suit the society. If any questions are raised by NGOs, they have to be answered and settled amicably. Unwanted scenarios to stop a Nano technology project should be avoided.

Indian Defense men aim to obtain cost effective solar and fuel cells with greater efficiency. Energy saving advantages is secured through Nano materials with LEDS and Nano catalysts to enhance combustion processes and good insulation materials.

These are suitable for army men.

This new technology should pave way for reduction in usage of fossil fuels and successfully develop renewable energy usage.

Elimination by Nano

Through Nano technology, the Indian scientists aim for water treatment and remediation. Waste water treatment, detection of virus germs in polluted water and purification will greatly help the soldiers toiling in hazardous areas along the borders. As water pollution is seen in many parts of Indian rivers. Nano technologies will positively eliminate water borne diseases and epidemic diseases. The army men, who use it, may be affected by poisonous inhalation and skin diseases.

The scientists raise doubts that items based on Nano technology may disrupt cellular, enzymatic and many unknown organ disorders to cause health problems to soldiers. They also feel that these Nano particles will be non-biodegradable and may develop new varieties of non-biodegradable pollutants and create deadly havoc to air, soil and water, while disposal. These risks are faced by army sector while production, packing and transportation to border areas. Special care is to be taken to dispose it off safely.

Life Cycle Analysis is to be done by Defense experts for all the instruments, devices and weapons using Nano technology. We lack sufficient data on EHS effects and LCA will be useful to identify and evaluate the expected risks for Nano technology applications and help to detect crucial knowledge gaps.

More info on Nano

We need more information on the nature and characteristics of applications of Nano materials. Suitable tests are to be conducted by R & D wing in Defense sector to save our soldiers from harmful health havocs. This needs the support of foreign experts in USA, UK and Israel with liberal fund allocation. Study in Regulatory Toxicology is also needed in an in depth manner.

Though many research works are done by local scientists they should join hands with global leaders in Nanotechnology to develop standardized protocols, reference materials and required database. It is important that denial of access to technology due to restrictive patent regime will lower country's capabilities to achieve success in it. Also all risk governance frame work should see that it includes all those who are affected by regulations, production and consumption of Nano.

Risk governance shall include the laws, process and institutions by which decisions based on risk analysis, communication and management are taken and carried out.

The Policy makers, Regulators, Businessmen, Scientific and Civil Society Communities, who are the vital players in risk governance frame work, should be strong. Their voices should be heard to ensure safety particularly for the defense men.

For success

Another point stressed by the defense chiefs is that production of Nano materials in large volumes is needed with consistency in quality and reasonable price tag. It is also important that Nano materials are supplied with proper particle size, surface chemistry, dispersion capability, compatibility etc to enable integration into the process.

Then only the Nano technology will be successful in Indian defense sector.

Advantages of Nanotechnology

Every sound heard by the enemy will enable them to detect our location in the war field. So, we have to equip our soldiers with sensors, which will emanate low noise at the same time send quick signals. This can be made possible only with nano structure.

The Physical and Chemical properties gained by Nanotechnology support the army men in the Warfield with comforts and user friendly environment. The small dimensions render the following advantages to warriors:

1) High functional density in defense is an important phenomenon gadget using nano electronics. It enhances the memory power and more data can be stored by the soldier to act with confidence. As he moves in the warfront with nano sensors and monitors he can send and receive more data to help his leader in the headquarters to take decisions. High density memory power is the back bone to take complete survey about the enemy movements and capture pictures of their camps hidden with ammunitions.

2) Function integration is made easy by small dimensions. This feature will support the sensing, DSP, radio, memory and power to get integrated. This is a boon to the warriors as he can comfortably handle the equipments, which have multifunctional capabilities at the same time small in size.

3) Efficiency is enhanced for the gadgets at any situation with speed. So, this will aid the army men for quick transportation with better optical and thermal characteristics.

4) Under any hazardous situations in forest, snow, mountains and flooded area, the soldier can easily handle the sensors and other military equipments meant for his security.

5) Of all other advantages, the military men can easily dispose the gadgets, which have miniaturization design.

The Nano technology provides materials, which have the following advantages:

1) As we have control at the nano scales, there is 100% possibility to obtain trouble free structures. This will add strength and conductivity to the defense equipments.

2) Large surface area is created by Nano structures and particles for betterment of sensing catalysis and absorption, when a soldier carries a sensor to detect the storage of detonators, explosives and hazardous chemicals. The small dimensional feature will enable him to detect sharply and correctly and smash the enemy attacks.

3) Totally new particles like carbon nano tubes can be created through nano level, which cannot be identified by others at war field. The terrorists can be trapped or misguided and thrown out with new and unidentifiable particles.

4) Night Vision Glasses and other gadgets with new optical effects can be produced with Nano structure to support the army men at border area and control trespassing activities.

5) Fluidic devices also play a vital role in defense for measurement and chemical processing deeds.

By minute and nano dimensions we get quick response to act swiftly by the defense personal at war field. Before the enemy can pounce, our soldiers can shatter their camps and ammunitions.

High throughput and lower chemical wastes are also added advantages seen in defense equipments with nano structure.

Single cell detection, Multi-parallel analysis and Matrix array will increase the efficiency for counter attacks.

Every sound heard by the enemy will enable him to detect our location in the war field. So, we have to equip our soldiers with sensors, which will emanate low noise at the same time send quick signals. This can be made possible only with nano structure.

The equipments carried by the soldiers can sense the movements of the enemies even at a great distance.

This is possible only with sensors having high sensitivity Radars, Missiles etc, which are equipped with Nano structure design for betterment of sensitivity.

****

International Conference on Condensed Matter and Applied Physics Bikaner 2017

ICC 2017 was a well organized international conference conducted in Engineering College Bikaner (Rajasthan State), which is an 18 years old institution on 200 acres land.

Interesting and informative topics on Condensed Matter &Applied Physics were presented and discussed by eminent scientists like Dr. Govind Prasad Kothiyal, former Head, Glass and Advanced Ceramic Division, BARC, Mumbai, Prof A.K.Nigam, TIFR, Mumbai, Dr.Chintamani Das, Program Officer (BSC) BARC, Mumbai, Prof.Michal Piasecki, Jan Dlugosz University, Poland, Prof.A.S.Patra, H.O.D.Department of Physics, Sidho-kanho-Birsha University, Purusia,West Bengal, Prof. Vijaya Srinivasa Vallabha Purapu, University of South Africa, Prof Govind Kothiyal Chairman, MRSI, Mumbai Chapter, Prof.Atiar Rahaman Molla, Senior Scientist, Prof Arghya Taraphder, Dept of Physics, IIT, Kharagpur,Prof.P.N.Gajjar,Dept of Physics, Gujarat University, Prof Mahesh Kumar, IIT, Jodhpur, Prof.CS.Sunandana, School of Physics, University of Hyderabad and Dr.M.S. Shekhawat, Dept of Physics, Govt Engineering College, Bikaner .

The Keynote Talk, Invited Talks and Contributory papers in this conference were the highlights.

Keynote session was chaired by Prof.A.K, Nigam, Tata Institute of Fundamental Research, Mumbai and co-chaired by Dr.Govind Prasad Kothiyal, BARC, Mumbai.

Prof A.K.Nigam vividly said about the TIFR, which functions under the control of DAE. He said that many popular research scholars, now submitting research papers all over the world in international conferences on physics, cosmic fireworks, hydrogen generator, quantum entanglement strange metals and black holes, predicting the properties of subatomic particles using large scale computer simulations, detecting of coronal explosion on the nearest planet – hostile star, etc. are from TIFR.

"Though major research works are conducted in Mumbai they have specialists engaged in scholarly works in Pune, Hyderabad and Bangalore. Branches of TIFR cover researches in Cell and Cancer Biology, Biological Physics, Fluorescence Spectroscopy Laser Spectroscopy, Molecular Genetics, NMR Spectroscopy and Molecular Bio Physics, etc said he and also invited more number of students in India and abroad to join their research team and courses offered in Physics, Chemistry and Mathematics etc.

He added that TIFIR is also engaged in conducting informative conferences like Conference on Advances in Catalysis for Energy and Environment 2018, under the leadership of Prof Vivek.Polshettiwar, and S.Mazumdar.

Dr. Govind Prasad Kothiyal, ex-head of Glass and Advanced Ceramic Division, BARC said that vital R & D works are carried out in Material Science and Engineering Division of BARC like characterization of alloys, their selection and design activities, ageing management, study about corrosion behavior of materials used in Nuclear Power Plants.

Developments of new methods are also carried out for the synthesis and preparation of enriched Boron Carbide Pellets for control rod applications and fission type Neutron Sensors with enriched Uranium coatings.

Studies in amorphous and nano materials, beam processing of materials, advanced techniques for joining of materials, materials for fuel cells and energy conversion materials are also conducted by Indian Scientists, spoke he.

2 sub sessions, which followed the keynote covered the topic Chalgogenide Compounds as promised multifunctional optoelectronic materials for IR spectral range and low field tunable micro wave absorption in magnetic, multiferroic and super conducting materials.

For the special session meant for invited speakers, eminent scholars like Prof Michal Piasecki, from Poland Jan Dlugosz University, Dr.Chintamani Das, Program Officer (BSC) BARC, Mumbai, Prof.P.N.Gajjar Department of Physics, Gujarat Prof.A.S.Patra, from Sidho Kanho Birsha University, Purulla in West Bengal Prof Shiv Prasad, IIT, Mumbai and Prof Ananthakrishnan Srinivasan, IIT, Guwhati were there to make it an intellectual meet.

Poster sessions on both days attracted several delegates and scientists in this event sponsored by Science and Engineering Research Board, Vasant Square Mall, Pocket-7, Sector B, VasantKunj, New Delhi – 70.

This popular research board functions under the Department of Science and Technology, Government of India, which has national and international partnership programs for graduate research opportunities worldwide program, overseas doctoral fellowship and partnerships for international research and education.

This great technical board is headed by Dr.Rajiv Sharma, Secretary, G grade Scientist, Head of Technology Missions Divisions of Dept of Science and Technology, which monitors and executes Nano Science and Technology mission and Clean Energy Research Institute.

ICC 2017 created an intellectual atmosphere with the efforts of organizing committee members comprising of Prof.Dinesh Shringi, Principal Government Engineering College, Bikaner, Dr.S.K.Bansal, Principal Govt College of Engineering & Technology, Bikaner, Dr.M.S.Shekhwat, Govt of Engineering & Technology, Bikaner, Dr.Sudhir Bhardwaj, Govt College of Engineering & Technology, Bikaner and Dr.Bhuvaneshwar Suthar, Convener ICC 2017. Their works covering paper presentation, local hospitality and publications were coordinated by Dr.Alka Swami,

GCET, Dr.Avinash Daga, Mrs.Garima Prajapat, Gourav Joshi, Laxman Singh, Dr.Mahendra Vyas, Dr.Naveen Sharma, Mr.Pankaj Jain, Dr.Praveen Purohit, Dr. Preeti Naruka, Ramachandra Beniwal, Dr.Ruma Bhadoria, Dr. Shivangi Bissa, Dr.Shivkumar, Dr. Shoukat Ali, Dr.Suresh Purohit, Dr.Vijay Makar, Dr.VijaySharma and Dr.Y.N.Singh.

The event organizers relaxed the delegates, participants and students with cultural program. A beautiful native girl gave a mesmerizing belly dance with five fire pots and 25 big pots on her head. She danced with balance on nails, bottles and broken glasses. It was actually a refreshing time for one and all. Some participants also took part in it to entertain the audience.

The first floor gallery accommodated the poster presentations on various topics like photonic materials, plasmonics, super conductivity magnetism and spintronics, nano materials, computational methods and applied physics, glasses & ceramics, composites, single crystals and novel materials etc.

Train tickets, hotel rooms and sightseeing were arranged for the participants through college volunteers to coordinate. Their hospitality was delightful for the visitors, who have come from various locations like Andhrapradesh, Tamilnadu, Gujarat, Bangalore, Kerala, West Bengal, Mumbai, Kharaghpur, Kolkata etc and foreign delegates from USA, U.K, South Africa, Italy and Poland Malaysia etc.

Students were from IIT, IISC (Bangalore) JNC, Jeppiar University, SRM College, SASTRA, Tamil Nadu, College of Engineering Guindy, TNadu, Bharathidasan University, Officials from DRDO,BARC, Ministry for Science & Technology, DAE, ARDE, Biotechnology experts and Research students from Cell Tower and Mobile Radiation Program, Impacting Research Innovation and Technology, Prime Ministers Fellowship for Doctoral Research Program, Banaras University, Professors and Research experts from Amity University (U.P), Jawaharlal Nehru Centre for Advanced Scientific Research, Bangalore, School of Nano Science and Technology, Calicut, Amrita Center for Nano Science and Technology, Calicut, Amrita Center for Nano Science, Center for Nano Technology and Advanced Bio Materials, Tanjore Tamilnadu, Department of Physics, IISC Bangalore, Centre for Research in Nano Technology and Science, Mumbai, Department of Mechanical Engineering and Materials Science Program, Kanpur, National Institute of Technology, Haryana and IIT (Delhi, Kharaghpur, Chennai).

Conclusion:

To develop Nano technology in India, such international conferences are a must. It was learnt from the organizer that there was overwhelming response and many were eager to attend but could not be accommodated. So, suitable auditorium or vast ground is to be selected in future to attract more students, technocrats and industrialists in this field to participate and get benefitted technically and economically.

As said by a DRDO official "There are myriad of applications of nano materials in every walk of life today and there is a need is to bring together all stake holders together to fructify the benefits of this technology. There is also a need for concerted efforts of R & D organizations academia and industry to realize this goal. Such conferences are the right platforms to generate innovative ideas and exchange of technical views to further the cause".

Let many more such events take place in all States of India.

****

Exhibitors in NSNST 2017

The intellectual symposium on Nano Science and Nano Technology gave an opportunity to the delegates to view some of the industrial contributors in the field of Nano.

The i2n Technologies located in Indian Institute of Science premises has the advantage of using sophisticated facilities available in Centre for Nano Science and Engineering. They serve for enlightening nano technology for startups with consultancy for infrastructure development, nano fabrication, training in nano equipments, gadgets and devices. They also undertake maintenance of Nano Labs and many educational institutions avail their services for establishing Nano Research Centers economically.

Atomic Force Microscope and Scanning Tunneling Microscope produced by them were displayed to the visitors.

Under the leadership of Ajaysingh, Rudra Pratap and Shiwani Singh this i2nTechnologies has fabricated Glider Kit, High Voltage Amplifier and Tip Etching Station. Contact address: 2nd floor, Entrepreneurships Center, STD Indian Institute of Science, Bangalore-560012. Ph: 080 – 23603046.

Shanmukha Innovation P ltd was another exhibitor in the symposium, which displayed healthcare and environmental based products fabricated by them.

Malaria Diagnostic Device, Nanorice making Device and Milk Adulterants Detection Device created by them was displayed in their stall and they were admired by the visitors.

They have specialized in developing Novel Opto fluidic technology based products. Water and soil quality testing sensors are developed with the support of CeNSE specialists.

Contact Cell No: 9008974499. Contact Address: 101, KK Towers, Balaji Layout, Kaggadasapura, C.V.Raman Nagar, Bangalore – 560093. Under the Directorship of Sai Siva Gorthi, Udaya Bhasker Sivarama Varnasi and Sairam Chinta it is functioning well with innovations in Nano-Technology.

Centre for Nano Science and Engineering had a stall to display their facilities given to researchers. The Ministry of Communication and Information Technologies, Government of India undertaking provides financial support to INUP functioning for the past 9 years in IISC campus to encourage both Indian and International research scholars in Nano Science. Details about workshop lectures, tutorials and hands-on-training are given to assist talented scientists. After the training they are permitted to execute the research works.

Sensors designed and supplied to military and civil aircrafts were displayed to the visitors. They were seen with a range varying from 0-1.2 bar, 0-600 mbar, 0-10bar, 0-600mbar and 0-150mbar. These sensors are fitted in engines, hydraulics system, environmental measuring systems and helicopters.

Their Pressure Sensors, Gas Analyzers, Devices used for respiration technology, anesthesia, environmental and climate protection hydraulic and pneumatic units and ventilation systems were seen in the show.

Visitors from Srilanka, Malaysia, UAE, Bangladesh, UK and Singapore were keenly interested in their workshop services given to scientists.

Their products and inventions suitable for mobile phones, cycling computers and cameras with altimeter capabilities were attractive.

Their pressure sensor is used for pedestrian protection, fuel tank vapor pressure monitor, engine management system, power train electronics and soot particle filters. Indian Air Force engineers and Army officials enquired about the significance of these devices and requested for more details with catalogs.

The exhibitors received good response from within India and from abroad researchers and manufacturers engaged in Nano field in this symposium.

Indian Scientists and Trade Fairs on Nano Technology

The Laser and Nano Technology Conferences conducted by IEEE usually receive rousing response all over the world. So, the Indian Nano and Laser experts promptly attend these conferences to update their knowledge.

Indian Institutes and the government encourage the Event Managers and leading Universities to conduct International Conferences, Seminars and Workshops on Nano Technology.

Indian Laser Association in Indore invites foreign delegates and scientists for the events they conduct in popular Universities like Sri Venkateswara University, Tirupati and Andhra Pradesh.

Eminent scientists like A.S. Joshi, Laser Plasma Division, Raja Ramana Centre for Advanced Technology (RRCAT); Dr. C.K. Jayasankar, Department of Physics in Sri Venkateswara University, Tirupathi; Mr. Viraj P. Bhanage of Laser Electronics Support Division in RRCAT, Indore; Dr.K.T.R.Reddy, Department of Physics, S.V.University, Tirupati; Dr.K.S.Bindra, RRCAT, Indore; Rajiv Jain, Laser Electronics Division, RRCAT, Indore and members of RRCAT join their hands to participate and present papers in Laser Technology meet in India and abroad.

The topics for which they show interest are Mode Locked Fiber Lasers, Fiber Bragg Gratings. Optical Fiber Sensors, Yb Doped CW Fiber Laser, various techniques of Rare Earth Doped Fibers, Double Clad Fiber Architecture, Terahertz Techniques and Applications, Accelerator Based Sources Terahertz Generation, Terahertz Optics and Materials, Femtosecond Laser Based Terahertz Generation and Detection Techniques, THz Imaging and Applications, Terahertz Transient Spectroscopy and Applications etc.

These scientists attend symposiums conducted by Optical Society of India, IEEE Lasers and Electro Optics Society, Power Beam Society of India and Photonics Society of India. These experts attend discussions and interactions arranged with foreign experts to update their knowledge on Nano Technology and Laser Science.

They attend Trade Fairs to deliver lectures on topics like High Power Laser, Beam Control Technologies, Thin Film Coatings, Low Energy Solid State Lasers, Combustion Driven High Power Gas Dynamic Laser, Nd-YAG Laser Designator cum Range Finder, Electro Optic Counter Measurements, Laser Related Optics, Laser Crystal, Chemical Oxy-Iodine Laser etc.,

The Laser Conferences conducted by IEEE usually receive rousing response all over the world. So, the Indian Nano and Laser experts promptly attend these Conferences. As they can meet foreign experts in these Conferences, they seek permission from their respective Universities to participate. The Institutes and Universities will bear all the expenses for them to attend both in India and abroad. Those who contribute papers to IEEE Journal of Quantum Electronics and Journal of Light Wave Technology participate in these Technical Congress Sessions.

Prof P. Basu, Institute of Radio Physics and Electronics and Dr. A.K.Gupta from DOE, who is the Chairman of Joint IEEE Chapter of Aerospace of Electronics Systems Society and Communications Society and Laser and Electro Optics Society participate and organize their informative sessions. These Nano and Laser specialists also attend discussions with foreign scientists held by the network of researchers, educationalists and entrepreneurs in Laser field. The Photonic Society of India in Cochin arranges this meet in their campus.

Leading researchers also take part in conferences conducted by Power Beam Society of India to promote industrial applications with this new Laser and Nano Technology.

Dr. M. Premasundaram, a popular Scientist G from LASTEC, obtained Fellowship from an U.K University in Laser Technology. He is well admired for his technical papers submitted on Solid State of Lasers and Combustion Driven Gas Dynamic Lasers. He attends conferences on High Power Laser and its Applications.

Dr. A.L. Dewar, a leading scientist retired from LASTEC, attends conferences covering Integrated Optics and Thin Films, High Power Chemical Oxy-iodine Lasers, Semi Conducting Transparent Thin Film etc. He has done his research works in University of New Castle and University of Glasgow. Hence he attends seminars conducted abroad frequently.

Prof. Arindam Ghosh from IISC is a reputed Young Scientist in India. He gives Power Point Presentations in leading conferences and seminars on Nano Science. He has even won the Oxford Instruments Young Nano Scientist Award 2015.

Other leading scientists in India, who actively participate in all seminars on Nano Technology, are Prof C.N.R Rao (JNCASR), Prof. A.K. Ganguly (Director INST), Prof Ashutosh Sharma (IIT Kanpur) and Prof K.N.Ganesh (Director IISER Pune).

Indian Scientists and foreign delegates meet in SASTRA University, Tanjore in Tamilnadu during Symposiums on Nano Technology and Advanced Bio Materials. The University has a large auditorium with excellent infrastructure to arrange lectures on Nano Sensors, Nano Medicine, Nano Devices, Nano Electronics and Nano Materials. Popular scientists like Dr.C.N.R. Rao and Prof A.K.Sood will participate in such symposiums.

Indian Nano experts and scientists take interest to attend conferences on Nano Fabrication Technologies, Translation Nano Medicine and Scanning Probe Microscopy, wherever it is held. When such seminar is arranged in foreign countries, the concerned Institutes will sponsor them and also encourage them to submit papers.

JNCASR has deputed several research scholars to various conferences to gain wide knowledge in Nano Technology. The prestigious Bharat Ratna award winner Prof C.N.R. Rao guides many scientists to present papers in international seminars.

Research experts in this field are Prof.P.S.AnilKumar, Department of Physics, Indian Institute of Science, Bangalore; Prof.Ashok Kumar Ganguly, Department of Chemistry, I.I.T Newdelhi; Prof.M.S.Bobji, Department of Mechanical Engineering, Indian Institute of Science, Bangalore; Dr.C.V.Dharmadhikari, Indian Institute of Science and Education, Pune; Prof.Ashok. M.Raichur, Department of Materials Engineering, IISC Bangalore; Dr.Praveen Kumar Vermula, Institute of Stem Cell & Biology & Regeneration Medicine, NCBS, Bangalore; Prof.Upadrasta Ramamurthy, Department of Materials Engineering IISC, Bangalore; Dr.Sachin Parashar, Advanced Materials & Green Chemistry, Tata Chemicals Ltd, Pune; Dr.Priyanka, Institute of Nano Science & Technology, Mohali; Dr.Subramian Krishnakumar, M.D, Vision Research Foundation, Chennai; Prof.Soumya Mukherji, Department of Bio Science & Bio Engineering, Biomedical Engineering IIT, Mumbai; Prof.Navakanta Bhat, Centre for Nano Science & Engineering (CEWSE); Department of Electrical Communication Engineering IISc; Dr.A.Sreekumaran Nair, Amrita Centre for Nano Science & Molecular Medicine, Amrita Institute of Medical Sciences, Kochi; Dr.S.S.V.Ramkumar, Automotive Oils & Nano Tech, Indian Oil Corporation, Faridabad and Dr.M.V.Reddy, Department of Materials Science & Engineering and Physics, National University of Singapore.

Other leading specialists interested in participation of technical conferences on Nano and Laser Technology are Dr.Praveen Asthana, Nano Mission Department of Science & Technology, Govt of India ; R.Rajan, United Nanotech Innovations P Ltd, Bangalore; Gadhadar Reddy, NoPo Nano Technologies P Ltd, Bangalore; Dr.C.Vankatesh, i2n Technologies P Ltd, Bangalore; Prof.Rudra Pratap, Centre for Nano Science & Engineering (CeNSE) Indian Institute of Science, Bangalore; Prof. G.U.Kulkarni, Chemistry and Physics of Materials Unit, JNCASR, Bangalore; Prof.Krishna N.Ganesh, Indian Institute of Science Education and Research, Pune; Prof.D.D.Sharma, Solid State and Structural Chemistry Unit, IISc Bangalore; Dr.Shantikumar V.Nair, Amrita Vishwa Vidyapeetham (University) Amrita Centre for Nano Science & Molecular Medicine, Kochi, etc.

These Scientists in India are deeply involved in Nano technological research projects and seminars concerned with Satellites. They visit Fairs and Congresses held abroad and present papers. All organizers like FICCI, IEE, JNCASR and IIT Institutes etc, invite them to give lectures and presentations in seminars.

Nano Materials for Checking Global Warming

Nano materials and Nano technology not only paves way for betterment of current technologies but also lowers the usage of fossil fuels. Its effort to reduce emission in vehicles by minimizing their weight and also decreasing fuel consumption will have a telling effect on combating global warming.

Entire world is focusing on global warming issues. Cities act as the source of Co2 (Carbon dioxide), which has a dangerous link to climate changes. This change is considered to be the greatest challenge facing the society. "Warming of the climate system is unequivocal, human-linked and various unwanted changes have been witnessed in climate system since 1980" say the environmental experts.

Controlling the climate change will need substantial and sustained reductions of greenhouse gas emissions.

While the developing and newly industrialized nations improve their standards of living, their usage of air-conditioning and other weather dependant consumption will likely enhance their reaction to climate change. At the same time reducing consumption and maintaining more sustainable life styles in developed countries will likely represent the best solution to reduce carbon emissions.

Energy efficiency is an ideal path to manage and restrain the growth of energy consumption. It is observed to be the easiest and most cost effective method to fight against climate changes.

Important strategies to be followed to combat global warming are:

1) Lowering energy consumption by employing more modernized technologies that minimize the usage of fossil fuels.

2) Adopting new technologies, which will utilize renewable energy and energy storage technologies.

3) Addressing carbon management that involves separation, capture, sequestration and conversion to valuable products.

The Climate Group has revealed that the extensive implementation of Smart Grid technologies will undoubtedly lower global emissions by 2.03 giga tonnes Co2e (Carbon dioxide equivalent is a quantity that describes for a given mixture and amount of green house gas, the amount of CO_2 that world have the same global warming potential, when measured over a specified time scale (say 100 years), worth €79 billion.

At this juncture, nano technology will play a vital role for smart grid and assist the reduction of CO_2 emissions. This technology not only paves way for betterment of current technologies but also lowers the usage of fossil fuels. Its effort to reduce emission in vehicles by minimizing their weight and also decreasing fuel consumption will have a telling effect on combating global warming.

For 10% reduction in weight of a vehicle we can achieve 10% reduction in fuel consumption and hence reduction in emissions.

So, Nanotechnology gives materials, which can lower the weight and also become stiffer and stronger.

By using Nanotechnology in aircraft coatings to protect the materials from environmental hazards, we can reduce weight and at the same time lower fuel consumption and CO_2 emission.

Nano coatings are significant in designing defense, marine equipments, plastic industries and automotives.

To enhance fuel efficiency, incorporation of nano catalysts is an ideal way. Third generation catalyst called enercat utilizes oxygen storing cerium oxide nano particles to complete fuel combustion and reduce fuel consumption.

"In automobiles it is important to reduce friction and increase resistance to wear and tear. By lowering the friction we can bring down fuel consumption by 2%"says a nano expert. This obviously drastically will cut down emission of 600 million tonnes of carbon dioxide per year through load carrying trucks and various types of heavy vehicles.

Silica is a valuable nano material, which reduces friction in vehicle tyres. Automobiles with Green Category A have proved to consume 7.5% less fuel than other vehicles with standard tyres.

It is found by environmental experts that 11% of total green house gas emissions are from commercial and residential buildings. Air conditioners and space heaters in buildings cover 40% of the total energy consumed by that building. Aerogel is an important nano structured item, which favorably limits the heat transfer through building elements and lowering heating loads on air conditioning and heating units.

Aerogel has the low density characteristics and a nano porous super insulating material. Also Silica Aerogel is well known as the lightest solid material with superior thermal insulating capabilities, high temperature stability, high surface area and very low dielectric constant.

Role of nanotechnology in smart grids is note worthy through nanotech sensors. It offers decentralized electricity monitoring abilities. Nano technology enhances battery power and also enable solar and wind power to render a higher share of overall electric power supply.

In photo voltaic also nanotechnology gives an ardent support to produce solar panels to increase the output.

With the support of graphene better performance is achieved in wind turbine, because one atom thick layer of mineral graphene is 100 times mightier than steel. Nanotechnology facilitates the light and stiff wind blade to spin at lower wind speeds, when compared to the regular blades.

Even in batteries, coating of nano particles on electrode increases the surface area and permit higher current flow between the electrode and the chemicals inside the battery.

Nanotechnology can be used for manufacturing wires with carbon nano tubes, which can carry greater loads and transmit power without losses for a stretch of several hundreds of kilometres. This will in turn enhance the power generating efficiency.

For economic development of a country the usage of highly energy intensive materials, such as steel, cement, glass and aluminium will increase. Such materials are vital for development of transport, housing, water management, infrastructure, construction and energy needs. Coal plays a major role in energy and energy intensive industries and infrastructure development activities.

Top 5 sources of CO_2 emissions are fossil fuel like coal, oil and natural gas, non energy use of fuels like erection works of infrastructure, iron and steel production, cement manufacturing and natural gas systems.

As per the statistics issued by the Netherlands Environmental Assessment agency it is said that since 2000, approximately 466 billion tons of CO_2 were emitted cumulatively by the human activities. Global CO_2 emission reached 30gt in the year 2010 and higher rate of 31.6gt in the year 2011.

It is said that 45% of CO_2 are emitted from the electricity generation sector by combustion of fossil fuel like coal, oil and natural gas to produce heat required to operate steam driven turbines followed by oil (35%) and natural gas (20%).

The World Resource Institute and International Energy Agency clearly states that CO_2 can be reduced by increasing the energy efficiency, conservation activities, greater reliance on renewable energy and usage of carbon capture and storage.

Smart Grid covers reliability, improved performance and resiliency with nanotechnology.

Reductions in CO_2 emission and usage of electricity are seen in Smart Grids.

Thanks to Nano Technology!

****

Nano Materials for Checking Global Warming

Nano materials and Nano technology not only paves way for betterment of current technologies but also lowers the usage of fossil fuels. Its effort to reduce emission in vehicles by minimizing their weight and also decreasing fuel consumption will have a telling effect on combating global warming.

Entire world is focusing on global warming issues. Cities act as the source of CO_2 (Carbon dioxide), which has a dangerous link to climate changes. This change is considered to be the greatest challenge facing the society. "Warming of the climate system is unequivocal, human-linked and various unwanted changes have been witnessed in climate system since 1980" say the environmental experts.

Controlling the climate change will need substantial and sustained reductions of greenhouse gas emissions.

While the developing and newly industrialized nations improve their standards of living, their usage of air-conditioning and other weather dependant consumption will likely enhance their reaction to climate change. At the same time reducing consumption and maintaining more sustainable life styles in developed countries will likely represent the best solution to reduce carbon emissions.

Energy efficiency is an ideal path to manage and restrain the growth of energy consumption. It is observed to be the easiest and most cost effective method to fight against climate changes.

Important strategies to be followed to combat global warming are:

1) Lowering energy consumption by employing more modernized technologies that minimize the usage of fossil fuels.

2) Adopting new technologies, which will utilize renewable energy and energy storage technologies.

3) Addressing carbon management that involves separation, capture, sequestration and conversion to valuable products.

The Climate Group has revealed that the extensive implementation of Smart Grid technologies will undoubtedly lower global emissions by 2.03 giga tonnes Co2e (Carbon dioxide equivalent is a quantity that describes for a given mixture and amount of green house gas, the amount of CO_2 that world have the same global warming potential, when measured over a specified time scale (say 100 years), worth €79 billion.

At this juncture, nano technology will play a vital role for smart grid and assist the reduction of CO_2 emissions. This technology not only paves way for betterment of current technologies but also lowers the usage of fossil fuels. Its effort to reduce emission in vehicles by minimizing their weight and also decreasing fuel consumption will have a telling effect on combating global warming.

For 10% reduction in weight of a vehicle we can achieve 10% reduction in fuel consumption and hence reduction in emissions.

So, Nanotechnology gives materials, which can lower the weight and also become stiffer and stronger.

By using Nanotechnology in aircraft coatings to protect the materials from environmental hazards, we can reduce weight and at the same time lower fuel consumption and CO_2 emission.

Nano coatings are significant in designing defense, marine equipments, plastic industries and automotives.

To enhance fuel efficiency, incorporation of nano catalysts is an ideal way. Third generation catalyst called enercat utilizes oxygen storing cerium oxide nano particles to complete fuel combustion and reduce fuel consumption.

"In automobiles it is important to reduce friction and increase resistance to wear and tear. By lowering the friction we can bring down fuel consumption by 2%"says a nano expert. This obviously drastically will cut down emission of 600 million tonnes of carbon dioxide per year through load carrying trucks and various types of heavy vehicles.

Silica is a valuable nano material, which reduces friction in vehicle tyres. Automobiles with Green Category A have proved to consume 7.5% less fuel than other vehicles with standard tyres.

It is found by environmental experts that 11% of total green house gas emissions are from commercial and residential buildings. Air conditioners and space heaters in buildings cover 40% of the total energy consumed by that building. Aerogel is an important nano structured item, which favorably limits the heat transfer through building elements and lowering heating loads on air conditioning and heating units.

Aerogel has the low density characteristics and a nano porous super insulating material. Also Silica Aerogel is well known as the lightest solid material with superior thermal insulating capabilities, high temperature stability, high surface area and very low dielectric constant.

Role of nanotechnology in smart grids is note worthy through nanotech sensors. It offers decentralized electricity monitoring abilities. Nano technology enhances battery power and also enable solar and wind power to render a higher share of overall electric power supply.

In photo voltaic also nanotechnology gives an ardent support to produce solar panels to increase the output.

With the support of graphene better performance is achieved in wind turbine, because one atom thick layer of mineral graphene is 100 times mightier than steel. Nanotechnology facilitates the light and stiff wind blade to spin at lower wind speeds, when compared to the regular blades.

Even in batteries, coating of nano particles on electrode increases the surface area and permit higher current flow between the electrode and the chemicals inside the battery.

Nanotechnology can be used for manufacturing wires with carbon nano tubes, which can carry greater loads and transmit power without losses for a stretch of several hundreds of kilometers. This will in turn enhance the power generating efficiency.

For economic development of a country the usage of highly energy intensive materials, such as steel, cement, glass and aluminium will increase. Such materials are vital for development of transport, housing, water management, infrastructure, construction and energy needs. Coal plays a major role in energy and energy intensive industries and infrastructure development activities.

Top 5 sources of CO_2 emissions are fossil fuel like coal, oil and natural gas, non energy use of fuels like erection works of infrastructure, iron and steel production, cement manufacturing and natural gas systems.

As per the statistics issued by the Netherlands Environmental Assessment agency it is said that since 2000, approximately 466 billion tons of CO_2 were emitted cumulatively by the human activities. Global CO_2 emission reached 30gt in the year 2010 and higher rate of 31.6gt in the year 2011.

It is said that 45% of CO_2 are emitted from the electricity generation sector by combustion of fossil fuel like coal, oil and natural gas to produce heat required to operate steam driven turbines followed by oil (35%) and natural gas (20%).

The World Resource Institute and International Energy Agency clearly states that CO_2 can be reduced by increasing the energy efficiency, conservation activities, greater reliance on renewable energy and usage of carbon capture and storage.

Smart Grid covers reliability, improved performance and resiliency with nanotechnology.

Reductions in CO_2 emission and usage of electricity are seen in Smart Grids.

Thanks to Nano Technology!

.